FOOTBALL SUPERSTARS ALBUM

1999

RICHARD J. BRENNER

AN EAST END PUBLISHING BOOK
SYOSSET, NY

Author's Note: All of the players in this book are gifted athletes, but they all had to work hard and overcome obstacles to achieve their success. Even a big star like Terrell Davis had to struggle to keep going through some very tough times. "I have always been in situations where things didn't look good," said Davis. "For a long time, I couldn't see the light at the end of the tunnel."

You can become a superstar, too, if you want to work for it. And there are lots of areas for you to consider besides athletics. You can help clean up the environment or work for world peace. You can become a musician, a teacher, a doctor, a human-rights activist, or a writer. We need superstar ecologists and caregivers more than we need superstar athletes.

Mr. Brenner, America's best-selling sportswriter, has written dozens of books, including: **BASEBALL SUPERSTARS ALBUM 1999, BASKETBALL SUPERSTARS ALBUM 1999,** and a series of easy-to-read photo-filled biographies of **MICHAEL JORDAN; GRANT HILL; KOBE BRYANT; MARK McGWIRE; and SAMMY SOSA.**

For complete details on how to order these and other exciting sports books, please see back page.

Mr. Brenner is also available to speak to student groups. For details, contact East End Publishing, Ltd., 54 Alexander Dr., Syosset, NY 11791 • (516) 364-6383.

To all people everywhere: May you always play in peace and happiness, walk in beauty, and work to build communities that are free from hate, fear, and bigotry of every type.

With great appreciation to everybody whose time and talents have contributed so much to this book, including, and most especially, Eric Macaluso; Peggy Hahn; Tony Inzerillo; Bob and Rob Tringali at SPORTSCHROME; and Mike Eby at ALLSPORT.
Thanks are also due, as usual, to Nell Fuqua at Scholastic Book Fairs.

Cover and interior design: **Eric Macaluso**

Photo Credits: The cover and interior photograph of Terrell Davis and the photograph of Jake Plummer were taken by **Tom Hauck**/ALLSPORT. Barry Sanders was taken by **Scott Halleran**/ALLSPORT. Dan Marino was taken by **Andy Lyons**/ALLSPORT. All of the remaining photos were supplied by SPORTSCHROME: The photographs of Vinny Testaverde, Drew Bledsoe, Mark Brunell, Doug Flutie, Antonio Freeman, Jamal Anderson, and Randall Cunningham were all taken by **Rob Tringali, Jr.** Garrison Hearst and Charles Woodson were taken by **Michael Zito**. Peyton Manning and Michael Strahan were taken by **Vincent Manniello**.

ISBN 0-943403-53-7

Published by EAST END PUBLISHING, Ltd.
54 Alexander Dr.
Syosset, NY 11791

Printed in the United States of America.

For many years, Native American groups have been appealing to sports teams not to use names and logos that they find offensive to their culture, such as "Redskins" or "Braves," or the laughing cartoon Indian on the uniform of the Cleveland baseball team. Any use of such names in this book will be extremely limited and for the purpose of identification only.

I urge all my readers to write to the offending teams and to the commissioners of Major League Baseball and the National Football League to protest the use of those names and logos.

- FINAL 1998 STANDINGS -

Team	W	L	T	PCT	PF	PA	Home	Road	AFC	NFC	DIV
AFC EAST											
N.Y. Jets	12	4	0	.750	416	266	7-1	5-3	10-2	2-2	7-1
Miami	10	6	0	.625	321	265	7-1	3-5	7-5	3-1	4-4
Buffalo	10	6	0	.625	400	333	6-2	4-4	7-5	3-1	4-4
New England	9	7	0	.563	337	329	6-2	3-5	7-5	2-2	4-4
Indianapolis	3	13	0	.188	310	444	3-5	0-8	3-9	0-4	1-7
AFC CENTRAL											
Jacksonville	11	5	0	.688	392	338	7-1	4-4	8-4	3-1	6-2
Tennessee	8	8	0	.500	330	320	3-5	5-3	7-5	1-3	7-1
Pittsburgh	7	9	0	.438	263	303	5-3	2-6	5-7	2-2	3-5
Baltimore	6	10	0	.375	269	335	4-4	2-6	5-7	1-3	2-6
Cincinnati	3	13	0	.188	268	452	1-7	2-6	2-10	1-3	2-6
AFC WEST											
Denver	14	2	0	.875	501	309	8-0	6-2	11-1	3-1	8-0
Oakland	8	8	0	.500	288	356	4-4	4-4	5-7	3-1	4-4
Seattle	8	8	0	.500	372	310	6-2	2-6	5-7	3-1	3-5
Kansas City	7	9	0	.438	327	363	5-3	2-6	4-8	3-1	4-4
San Diego	5	11	0	.313	241	342	4-4	1-7	4-8	1-3	1-7

Team	W	L	T	PCT	PF	PA	Home	Road	AFC	NFC	DIV
NFC EAST											
Dallas	10	6	0	.625	381	275	6-2	4-4	1-3	9-3	8-0
Arizona	9	7	0	.563	325	378	5-3	4-4	1-3	8-4	4-4
N.Y. Giants	8	8	0	.500	287	309	5-3	3-5	3-1	5-7	5-3
Washington	6	10	0	.375	319	421	4-4	2-6	2-2	4-8	2-6
Philadelphia	3	13	0	.188	161	344	3-5	0-8	0-4	3-9	1-7
NFC CENTRAL											
Minnesota	15	1	0	.938	556	296	8-0	7-1	4-0	11-1	7-1
Green Bay	11	5	0	.688	408	319	7-1	4-4	3-1	8-4	4-4
Tampa Bay	8	8	0	.500	314	295	6-2	2-6	2-2	6-6	4-4
Detroit	5	11	0	.313	306	378	4-4	1-7	1-3	4-8	4-4
Chicago	4	12	0	.250	276	368	3-5	1-7	2-2	2-10	1-7
NFC WEST											
Atlanta	14	2	0	.875	442	289	8-0	6-2	3-1	11-1	7-1
San Francisco	12	4	0	.750	479	328	8-0	4-4	2-2	10-2	7-1
New Orleans	6	10	0	.375	305	359	4-4	2-6	1-3	5-7	3-5
Carolina	4	12	0	.250	336	413	2-6	2-6	1-3	3-9	3-5
St. Louis	4	12	0	.250	285	378	2-6	2-6	3-1	1-11	0-8

WILD CARD GAMES

NFC
Arizona 20, DALLAS 7
SAN FRANCISCO 30, Green Bay 27

AFC
MIAMI 24, Buffalo 17
JACKSONVILLE 25, New England 10

DIVISIONAL PLAYOFFS

ATLANTA 20, San Francisco 18
MINNESOTA 41, Arizona 21

DENVER 38, Miami 3
N.Y. JETS 34, Jacksonville 24

CONFERENCE CHAMPIONSHIPS

Atlanta 30, MINNESOTA 27 (OT)

DENVER 23, N.Y. Jets 10

SUPER BOWL XXXIII

Denver 34, Atlanta 19
Sunday, January 31, 1999 • Pro Player Stadium, Miami, Florida

QUARTERBACKS

NAME	ATT.	COMP.	COMP%	YDs	YDs/ATT.	TDs	LONG	INT.	RATING
Vinny Testaverde, NYJ	421	259	61.5	3256	7.7	29	82t	7	101.6
John Elway, DEN	356	210	59.0	2806	7.9	22	58	10	93.0
Neil O'Donnell, CIN	343	212	61.8	2216	6.5	15	76t	4	90.2
Mark Brunell, JAC	354	208	58.8	2601	7.4	20	78t	9	89.9
Doug Flutie, BUF	354	202	57.1	2711	7.7	20	84t	11	87.4
Drew Bledsoe, NE	481	263	54.7	3633	7.6	20	86t	14	80.9
Rich Gannon, KC	354	206	58.2	2305	6.5	10	80t	6	80.1
Steve McNair, TEN	492	289	58.7	3228	6.6	15	47	10	80.1
Dan Marino, MIA	537	310	57.7	3497	6.5	23	61t	15	80.0
Warren Moon, SEA	258	145	56.2	1632	6.3	11	45	8	76.6
Jim Harbaugh, BAL	293	164	56.0	1839	6.3	12	66t	11	72.9
Peyton Manning, IND	575	326	56.7	3739	6.5	26	78t	28	71.2
Kordell Stewart, PIT	458	252	55.0	2560	5.6	11	55t	18	62.9
Donald Hollas, OAK	260	135	51.9	1754	6.8	10	47	16	60.6
Craig Whelihan, SD	320	149	46.6	1803	5.6	8	55	19	48.0
Ryan Leaf, SD	245	111	45.3	1289	5.3	2	67	15	39.0

RUSHERS

NAME	ATT.	YDs	YDs/ATT.	LONG	TDs
Terrell Davis, DEN	392	2008	5.1	70	21
Marshall Faulk, IND	324	1319	4.1	68t	6
Eddie George, TEN	348	1294	3.7	37t	5
Curtis Martin, NYJ	369	1287	3.5	60t	8
Ricky Watters, SEA	319	1239	3.9	39t	9
Fred Taylor, JAC	264	1223	4.6	77t	14
Jerome Bettis, PIT	316	1185	3.8	42	3
Corey Dillon, CIN	262	1130	4.3	66	4
Antowain Smith, BUF	300	1124	3.7	30	8
Robert Edwards, NE	291	1115	3.8	53	9

RECEIVERS

NAME	REC.	YDs	YDs/REC.	LONG	TDs
O.J. McDuffie, MIA	90	1050	11.7	61t	7
Marshall Faulk, IND	86	908	10.6	78t	4
Rod Smith, DEN	86	1222	14.2	58	6
Keyshawn Johnson, NYJ	83	1131	13.6	41t	10
Carl Pickens, CIN	82	1023	12.5	67t	5
Tim Brown, OAK	81	1012	12.5	49	9
Jimmy Smith, JAC	78	1182	15.2	72t	8
Wayne Chrebet, NYJ	75	1083	14.4	63t	8
Frank Wycheck, TEN	70	768	11.0	38	2
Ben Coates, NE	67	668	10.0	33	6
Eric Moulds, BUF	67	1368	20.4	84t	9

PUNTERS

NAME	PUNTS	YDs	LONG	YDs/PUNT
Craig Hentrich, TEN	69	3258	71	47.2
Tom Rouen, DEN	66	3097	76	46.9
Chris Gardocki, IND	79	3583	62	45.4
Bryan Barker, JAC	85	3824	65	45.0
Lee Johnson, CIN	69	3083	69	44.7
Tom Tupa, NE	74	3294	64	44.5
Jeff Feagles, SEA	81	3568	59	44.0
Darren Bennett, SD	95	4174	65	43.9
Kyle Richardson, BAL	90	3948	67	43.9
Josh Miller, PIT	81	3530	73	43.6
Leo Araguz, OAK	98	4256	64	43.4
Louie Aguiar, KC	75	3226	59	43.0
Klaus Wilmsmeyer, MIA	93	3949	57	42.5
Chris Mohr, BUF	69	2882	57	41.8
John Kidd, NYJ	41	1686	57	41.1

PUNT RETURNERS

NAME	RET.	YDs	YDs/RET.	LONG	TDs
Reggie Barlow, JAC	43	555	12.9	85t	1
Jermaine Lewis, BAL	32	405	12.7	87t	2
Terrell Buckley, MIA	29	354	12.2	35	0
Latario Rachal, SD	32	387	12.1	56	0
Desmond Howard, OAK	45	541	12.0	75t	2
Darrien Gordon, DEN	34	379	11.1	44	0
Joey Galloway, SEA	25	251	10.0	74t	2
Kevin Williams, BUF	37	369	10.0	73	0
Tamarick Vanover, KC	27	264	9.8	37	0
Damon Gibson, CIN	27	218	8.1	65t	1

KICKOFF RETURNERS

NAME	RET.	YDs	YDs/RET.	LONG	TDs
Corey Harris, BAL	35	965	27.6	95t	1
Steve Broussard, SEA	29	781	26.9	90t	1
Vaughn Hebron, DEN	46	1216	26.4	95t	1
Tremain Mack, CIN	45	1165	25.9	97t	1
John Avery, MIA	43	1085	25.2	55	0
David Dunn, PIT	21	525	25.0	44	0
Reggie Barlow, JAC	30	747	24.9	91	0
Aaron Glenn, NYJ	24	585	24.4	62	0
Derrick Cullors, NE	45	1085	24.1	68	0
Tamarick Vanover, KC	41	956	23.3	62	0
Ahman Green, SEA	27	620	23.0	57	0

TOUCHDOWNS

NAME	TDs	RUSHING	RECEIVING	RETURNS	POINTS
Terrell Davis, DEN	23	21	2	0	138
Fred Taylor, JAC	17	14	3	0	102
Robert Edwards, NE	12	9	3	0	72
Joey Galloway, SEA	12	0	10	2	72
Keyshawn Johnson, NYJ	11	1	10	0	66
Ed McCaffrey, DEN	10	0	10	0	62*
Marshall Faulk, IND	10	6	4	0	60
Shannon Sharpe, DEN	10	0	10	0	60
Ricky Watters, SEA	9	9	0	0	56*
Tim Brown, OAK	9	0	9	0	54
Curtis Martin, NYJ	9	8	1	0	54
Eric Moulds, BUF	9	0	9	0	54

5 tied with 8 touchdowns

** includes 2-point conversions*

KICKERS

NAME	PAT	ATT.	FG	ATT.	LONG	POINTS
Steve Christie, BUF	41	41	33	41	52	140
Al Del Greco, TEN	28	28	36	39	48	136
Jason Elam, DEN	58	58	23	27	63	127
Adam Vinatieri, NE	32	32	31	39	55	127
John Hall, NYJ	45	46	25	35	54	120
Pete Stoyanovich, KC	34	34	27	32	53	115
Mike Hollis, JAC	45	45	21	26	47	108
Mike Vanderjagt, IND	23	23	27	31	53	104
Norm Johnson, PIT	21	21	26	31	49	99
Olindo Mare, MIA	33	34	22	27	48	99
Todd Peterson, SEA	41	41	19	24	51	98
John Carney, SD	19	19	26	30	54	97
Matt Stover, BAL	24	24	21	28	48	87
Greg Davis, OAK	31	31	17	27	51	82

TEAM OFFENSE

NAME	RUSH YARDS	PASS YARDS	TOTAL YARDS	RUSH TDs	PASS TDs	RET.& REC.TDs	EP/ 2PT.CON	FG/ FGA	POINTS
Denver	2468	3808	6276	26	32	4	58/1	23/27	501
N.Y. Jets	1879	4032	5911	12	33	4	45/0	25/35	416
Buffalo	2161	3621	5782	13	28	2	41/0	33/41	400
Jacksonville	2102	3343	5445	19	24	4	45/1	21/26	392
Seattle	1626	3219	4845	11	21	13	41/2	19/24	372
New England	1480	4004	5484	9	23	3	32/1	31/39	337
Tennessee	1970	3482	5452	12	16	4	28/1	36/40	330
Kansas City	1548	3472	5020	19	15	1	34/0	27/32	327
Miami	1535	3582	5117	10	23	4	33/0	22/27	321
Indianapolis	1486	3739	5225	7	26	0	23/4	27/31	310
Oakland	1727	3534	5261	6	21	7	31/1	17/27	288
Baltimore	1629	3152	4781	7	16	6	24/2	21/28	269
Cincinnati	1639	3545	5184	7	20	4	21/2	19/27	268
Pittsburgh	2034	2781	4815	8	13	5	23/3	26/32	263
San Diego	1728	3115	4843	11	11	1	19/2	26/30	241

SACKS

NAME	SACKS
Michael Sinclair, SEA	16.5
Michael McCrary, BAL	14.5
Derrick Thomas, KC	12.0
Jason Gildon, PIT	11.0
Lance Johnstone, OAK	11.0
Trace Armstrong, MIA	10.5
Darrell Russell, OAK	10.0
Bruce Smith, BUF	10.0
Jason Taylor, MIA	9.0
Peter Boulware, BAL	8.5
Trevor Pryce, DEN	8.5
Maa Tanuvasa, DEN	8.5
Lorenzo Bromell, MIA	8.0
Ellis Johnson, IND	8.0

INTERCEPTIONS

NAME	INT.	YDs	LONG	TDs
Ty Law, NE	9	133	59t	1
Terrell Buckley, MIA	8	57	61	1
Sam Madison, MIA	8	114	35	0
Shawn Springs, SEA	7	142	56t	2
Aaron Glenn, NYJ	6	23	26	0
Greg Jackson, SD	6	50	25	0
Lawyer Milloy, NE	6	54	30t	1
Kurt Schulz, BUF	6	48	24	0
Rod Woodson, BAL	6	108	60t	2
Eric Allen, OAK	5	59	22	0
Duane Starks, BAL	5	3	2	0
Dewayne Washington, PIT	5	178	78t	2
Charles Woodson, OAK	5	18	46t	1
6 players tied with	4			

TEAM DEFENSE

NAME	RUSH YARDS	PASS YARDS	TOTAL YARDS	OPP. TDs	RUSH TDs	PASS TDs	RET. TDs	SACKS	INT.	POINTS ALLOWED
Miami	1511	2924	4435	28	6	17	5	45	29	265
N.Y. Jets	1659	3040	4699	30	11	16	3	36	21	266
Pittsburgh	1642	3321	4963	30	8	17	5	41	16	303
Denver	1287	3648	4935	38	8	28	2	47	19	309
Seattle	1999	3690	5689	35	13	18	4	53	24	310
Tennessee	1610	3511	5121	34	9	24	1	30	12	320
New England	1547	3635	5182	36	8	26	2	36	24	329
Buffalo	1493	3198	4691	39	11	27	1	43	18	333
Baltimore	1705	3592	5297	37	12	20	5	39	17	335
Jacksonville	2000	3559	5559	36	9	23	4	30	13	338
San Diego	1140	3068	4208	38	12	21	5	39	20	342
Oakland	1674	2876	4550	38	8	22	8	41	21	356
Kansas City	1869	2985	4854	44	22	17	5	40	13	363
Indianapolis	2570	3266	5836	52	20	27	5	38	8	444
Cincinnati	2612	3151	5763	53	23	23	7	28	13	452

OFFENSE

QUARTERBACKS

John Elway-x	Broncos
Vinny Testaverde	Jets
Doug Flutie	Bills

RUNNING BACKS

Terrell Davis-xy	Broncos
Marshall Faulk-x	Colts
Eddie George	Oilers
Curtis Martin-z	Jets

FULLBACK

Sam Gash	Bills

WIDE RECEIVERS

Ed McCaffrey-x	Broncos
Jimmy Smith-x	Jaguars
Keyshawn Johnson	Jets
Eric Moulds	Bills

TIGHT ENDS

Shannon Sharpe-xy	Broncos
Ben Coates	Patriots
Frank Wycheck-z	Oilers

TACKLES

Tony Boselli-x	Jaguars
Jonathan Ogden-x	Ravens
Tony Jones	Broncos

GUARDS

Bruce Matthews-xy	Oilers
Ruben Brown-x	Bills
Will Shields	Chiefs
Mark Schlereth-z	Broncos

CENTERS

Dermontti Dawson-x	Steelers
Tom Nalen	Broncos

DEFENSE

DEFENSIVE ENDS

Michael McCrary-x	Ravens
Bruce Smith-x	Bills
Michael Sinclair	Seahawks

INTERIOR LINEMEN

Tim Bowens-xy	Dolphins
Darrell Russell-x	Raiders
Ted Washington	Bills
Cortez Kennedy-z	Seahawks

OUTSIDE LINEBACKERS

Chad Brown-x	Seahawks
Mo Lewis-x	Jets
Bill Romanowski	Broncos
Peter Boulware-n	Ravens

INSIDE LINEBACKERS

Junior Seau-x	Chargers
Ray Lewis	Ravens

CORNERBACKS

Aaron Glenn-xy	Jets
Ty Law-x	Patriots
Shawn Springs	Seahawks
Charles Woodson-z	Raiders

STRONG SAFETIES

Rodney Harrison-x	Chargers
Lawyer Milloy	Patriots

FREE SAFETY

Steve Atwater-x	Broncos

SPECIALISTS

PUNTER

Craig Hentrich	Oilers

KICKER

Jason Elam	Broncos

KICK RETURNER

Jermaine Lewis	Ravens

SPECIAL TEAMS

Bennie Thompson	Ravens

NOTES: x - starter; y - injured; z - injury replacement;
n - "need player," selected by coach from
first alternates at defensive end, inside linebacker, or outside linebacker

TERRELL DAVIS

RUNNING BACK • DENVER BRONCOS

TERRELL DAVIS

DENVER BRONCOS

Born: October 28, 1972, in San Diego, CA
Height: 5′ 11″ **Weight:** 210
College: Georgia **Round drafted:** Sixth (1995)
NFL seasons: 5

CAREER STATS

Rushing				Receiving			
Att.	Yards	Avg.	TDs	No.	Yards	Avg.	TDs
1,343	6,413	4.8	56	152	1,181	7.8	5

Terrell Davis has set a singular pace during his first four years in the NFL, piling up yards and setting records as quickly as any running back ever has. "He'd go down as one of the greatest ever to play the game, even if he quit playing today," said Denver Broncos teammate Matt Lepsis. But Davis isn't overly interested in his place in football history at this point in his career. "I'm just running with my head down toward the finish line," he said. "When I get there, then I'll look back and see what happened."

In the meantime, Davis—who rushed for over 1,000 yards in his rookie year after entering the league as an unheralded sixth-round draft choice in 1995—is more interested in helping the Broncos win championships and in serving as an inspiration for other low-level draft picks. "I want to be able to play well so that people who are drafted in the late rounds won't be discouraged about trying to break into this league," said Davis, who wasn't expected to make it through his first training camp. "I want them to know that even if they are drafted later, they can still come in and contribute."

After contributing an AFC-high 1,538 rushing yards to the Denver cause in 1996, Davis had what appeared to be a career year in 1997, when he again led the AFC in rushing, with a whopping 1,750 yards. Then he continued galloping through the postseason, racking up 581 rushing yards and eight touchdowns in four games, including a three-touchdown, MVP-winning performance in Super Bowl XXXII. "This is Terrell's team now," said John Elway, Denver's future Hall of Fame quarterback. "He's the horse that carried us to the title."

But Davis turned the throttle still higher in 1998, as his powerful, slashing runs netted him 2,008 yards, the third-highest total in NFL history. "Shoot, Terrell could have rushed for 2,500 yards this season," said the Atlanta Falcons' defensive line coach Bill Kollar. "There were lots of games when Denver was up by so much that they sat him down way early."

Davis, who also led the league in first downs (110), rushing touchdowns (21), and total touchdowns (23, the third-highest number in NFL annals), was a runaway choice as the league's MVP. "He's the only player I've ever been around who doesn't have a weakness," said Broncos' head coach Mike Shanahan. "He has everything you look for in a running back. He's extremely smart and dedicated. He runs with elusiveness and explosiveness, and he has the speed to make the long run. He can pass protect, and he can block in short yardage. He's the most complete back I've ever seen."

Davis, who has rushed for 105.1 yards per game, the highest average in NFL history, may not be ready to think about his place in football history, but he knows where he wants to be at the end of his career. "When I hang up my cleats for the last time, I would like to think that I'm the greatest back to play the game," said Davis. "But I still have a lot of work to do, games to win, and—hopefully—Super Bowls to win. If you want to put your signature on the game, that's the bottom line."

DREW BLEDSOE

QUARTERBACK • NEW ENGLAND PATRIOTS

DREW BLEDSOE

NEW ENGLAND PATRIOTS

Born: February 14, 1972, in Ellensburg, WA
Height: 6′ 5″ **Weight:** 233
College: Washington State **Round drafted:** First (1993)
NFL seasons: 6

CAREER STATS

Att.	Comp.	Pct.	Yards	TDs	Int.
3,382	1,887	56.0	21,981	128	102

Although Drew Bledsoe has had doubters at every level at which he's ever played, he always seems to find a way to emerge on top of the pile. "He was just a gangly kid who had big feet and was a little bit clumsy," recalled Shorty Bennett, who was Bledsoe's coach at Walla Walla High School. But he overcame those problems and was the team's starting quarterback by the end of his sophomore year. By the end of his senior season, Bledsoe, fully grown into his large feet, threw for 2,560 yards and 25 touchdowns, and was named the 1989 Washington State High School Player of the Year. Then he moved on to Washington State University, where he became a three-year starter, progressing so rapidly that he was taken as the number one overall pick in the 1993 NFL draft. "Over the years," noted Bennett, "he's demonstrated a talent that allows him to exceed all expectations of what he can do."

Although common wisdom holds that young quarterbacks should be brought along slowly, Bill Parcells, who was the head coach of the New England Patriots, created a trial by fire by naming Bledsoe as his opening-day starter. Although the Pats were short of talent at the time, Bledsoe responded to the pressure by passing for 2,494 yards, the fourth-highest rookie total of all time. He followed that sound first-year effort with a truly spectacular second season, in which he led the league with 691 attempts (a new NFL record), 400 completions (four shy of the all-time mark), and 4,555 passing yards, while becoming the youngest quarterback ever to play in a Pro Bowl game. Even more important, Bledsoe, who was twice named AFC Player of the Week, passed the Pats out of the basement and into the postseason for the first time in eight years.

During his six seasons in Foxboro, Bledsoe has gone on to become one of the most prolific passers in NFL history, while leading the Patriots to four playoff appearances, a pair of divisional titles, and a trip to Super Bowl XXXI. The youngest quarterback to throw for 10,000 yards, Bledsoe is the second-fastest to pass for 20,000 yards, achieving that milestone in his eighty-third game, nine more than the Miami Dolphins' Dan Marino needed but three fewer than the Green Bay Packers' three-time MVP Brett Favre. Bledsoe finally seemed to win over his few remaining critics last season, when, playing with a fractured finger on his throwing hand, he became the first and only NFL quarterback ever to complete game-winning touchdown passes in the final 30 seconds in two consecutive games. "I enjoy those situations," said Bledsoe. "It's the situation you live for when you play sports, from the time you're playing ball at recess in grade school."

"Maybe he's finally making people realize what they should have realized a few years ago," said football analyst and former NFL signal caller Phil Simms. "They should have seen that he's just an outstanding quarterback. He's always been in my top five. He is absolutely a big-time veteran who is still young, and that makes him a rare commodity."

DOUG FLUTIE

QUARTERBACK • BUFFALO BILLS

DOUG FLUTIE

BUFFALO BILLS

Born: October 23, 1962, in Manchester, MD
Height: 5' 10" **Weight:** 175
College: Boston College **Round drafted:** Eleventh (1985)
NFL seasons: 5

CAREER STATS

Att.	Comp.	Pct.	Yards	TDs	Int.
695	368	52.9	4,914	34	27

Although Doug Flutie has excelled at each level he's played at, critics have continued to downplay his ability because of his relatively small size. Even after Flutie had become a three-sport star at Natick (Massachusetts) High School, most college coaches and advisers told Flutie that he was too short to play big-time college football. "Height?" snorted Flutie, dismissing the idea of measuring people with a ruler rather than by their accomplishments. "I never even thought about my height."

Flutie did finally receive a scholarship from Boston College, but when he showed up for practice he discovered that the coaching staff had already decided that he was too small to play quarterback and had him listed on the depth chart as a wide receiver. "I had convinced myself that I wasn't going to be a quarterback," recalled Flutie. "It didn't matter, as long as I could play major college football."

But in the fourth game of his freshman year, a combination of injuries and ineffectiveness by the Eagles' three signal callers caused coach Jack Bicknell to turn to Flutie, more in desperation than expectation. "It was like somebody hit a switch and picked up the tempo of the entire team," recalled Bicknell. "Never, ever did we imagine what we had in Doug."

As it turned out, what they had was a four-year starter whose ability to deliver pinpoint passes and scramble out of trouble led the Eagles to three straight Bowl appearances. Flutie also connected on one of the most famous touchdowns in football history, a last-second "Hail Mary" that lifted BC to an improbable 47–45 win over the University of Miami. The play, one of the most frequently shown sports highlights of all time, is known to football fans everywhere as The Pass. By the time he finished his fabled college career, Flutie had captured the 1984 Heisman Trophy and thrown for more yards (10,579), and rolled up more total offense (11,054) than any player in the history of college football.

When those credentials didn't seem to impress NFL teams, Flutie played a year in the USFL—a short-lived rival pro league—before beginning a four-year stint in the NFL, first with the Chicago Bears and then with the New England Patriots. But after getting only 14 starts in four seasons, Flutie decided to take his game up north, to the Canadian Football League. "They didn't want me down here," said Flutie, who was named the CFL's MVP six times in eight seasons. "That frustrated me, because I knew I had the skills to play. But I never really got the chance."

Last year, when Flutie finally did get his chance, he showed that he really did have NFL-level skills by leading the Buffalo Bills into the playoffs while earning a berth in the Pro Bowl for himself. "Nobody can tell you what you can and can't accomplish in life; it's limitless," declared Flutie. "I had people telling me that I couldn't play football on a Division 1 level, and that I couldn't make it in the NFL. I've found out that you have to make decisions for yourself, know what your capabilities are, and take some positive chances with your life."

MARK BRUNELL

QUARTERBACK • JACKSONVILLE JAGUARS

MARK BRUNELL

JACKSONVILLE JAGUARS

Born: September 17, 1970, in Los Angeles, CA
Height: 6′ 1″ **Weight:** 218
College: Washington **Round drafted:** Fifth (1993)
NFL seasons: 6

CAREER STATS

Att.	Comp.	Pct.	Yards	TDs	Int.
1,719	1,038	60.4	12,512	72	43

While some boys begin dreaming of playing in the NFL from the time they throw their first Nerf ball spiral, Mark Brunell has always had a myopic view of his football development. "I've just taken it one step at a time," says Brunell, talking about a career that stretches back to his childhood in Santa Maria, California.

Although Brunell had greater natural ability as a baseball player, he lost his heart to football in his sophomore year at Saint Josephs High School. "Football was more exciting, more intense," said Brunell, who was selected as one of the top 100 prep players in the nation in 1987 by *The Sporting News*. "It was more of a challenge. I just started enjoying football more than baseball."

Brunell, a shifty runner and accurate passer as well as a top student, worked diligently to develop his skills and build up his body. "Mark had a mental toughness and discipline that I had never seen before," said John Morris, who was Brunell's assistant baseball coach. The combination of skills and dedication to excellence earned Brunell a scholarship to the University of Washington, where he developed into one of the top-tier college quarterbacks. The left-handed scrambler, who was named the MVP of the 1991 Rose Bowl after he'd thrown for a pair of scores and run for two more, was eventually snapped up by the Green Bay Packers in the fifth round of the 1993 draft. Although he received only scant playing time while backing up Brett Favre, he made a powerful impression during his two-year stay in Green Bay. "He's simply a defensive back's nightmare, because you never know if he's going to run or throw," said LeRoy Butler, the Packers' All-Pro strong safety, who used to practice against Brunell. "But if he's on your side you know that one way or the other he'll keep you in the game."

Tom Coughlin, the head coach of the Jaguars, shared Butler's view of Brunell, and worked out a trade that brought Brunell to Jacksonville in the spring of 1995, just prior to the start of the Jags' initial season. "I was excited about Mark's athleticism, his leadership, toughness, arm strength, and ability to move in the pocket," said Coughlin. "He was the one I wanted."

Brunell began justifying Coughlin's judgment from the get-go, finishing the season as the highest-rated passer ever for a first-year expansion team. The strong-armed southpaw led all quarterbacks in both rushing and passing the following season, something which hadn't been done since 1963, when Hall of Famer Johnny Unitas turned the trick. Brunell's electrifying play also sparked the Jags into the postseason, where they zapped the Buffalo Bills and the heavily favored Denver Broncos. "He just made great plays all day," said John Elway, who would lead Denver to Super Bowls in each of the next two seasons. "You don't see a lot of guys who can make things happen like he can."

Brunell has continued to play at such a high level that many knowledgeable fans predict that he's only one step away from the ultimate destination—the Super Bowl. "Mark is a brilliant quarterback," said former Kansas City Chiefs' coach Marty Schottenheimer, explaining that he doesn't use the word lightly. "I think he's the next great quarterback."

VINNY TESTAVERDE

QUARTERBACK • NEW YORK JETS

VINNY TESTAVERDE

NEW YORK JETS

Born: November 13, 1963, in Brooklyn, NY
Height: 6′ 5″ **Weight:** 238
College: Miami **Round drafted:** First (1987)
NFL seasons: 12

CAREER STATS					
Att.	Comp.	Pct.	Yards	TDs	Int.
4,598	2559	55.7	32,479	204	190

Vinny Testaverde was riding high after the Tampa Bay Buccaneers made him the top overall pick of the 1987 draft. "I had won the Heisman, all kinds of awards; life was great," said Testaverde, who had quarterbacked the University of Miami to a 21–1 record in his two seasons as the Hurricanes' starter. "I was the number one pick in the draft. I had signed for a lot of money. I didn't know what was waiting for me or how things could turn around so quickly. But going to Tampa Bay was like getting smacked in the mouth by a two-by-four."

Unfortunately for Testaverde, he played for the Buccaneers during a period when they were the worst-run and least-talented franchise in the NFL. "There was a lot of pressure put on him," recalled Sam Wyche, who was one of the three head coaches that Testaverde played under in his six straight losing seasons in Tampa Bay. "He was supposed to come in and save the Buccaneers, and no one player can do that."

His stay in Tampa Bay quickly turned into a horror show for Testaverde, who threw 35 interceptions in 1988, the second-highest total in NFL history. By the time his tour of duty ended, following the 1992 season, he had thrown a total of 112 picks and only 77 touchdowns, and his confidence and reputation were on a steep decline.

Testaverde spent the following three seasons in Cleveland and then, after the Browns morphed into the Ravens, another two in Baltimore. Surrounded by a higher level of talent than he had enjoyed in Tampa Bay, Testaverde finally enjoyed his first winning season in 1994, and two years later he looked as if he had regained his Heisman form, as he threw for 4,177 yards and 33 touchdowns. Although his stats were strong enough to earn him his first trip to the Pro Bowl, the Ravens won only four games, and after the team tanked again in 1997, Testaverde was released.

After 11 mostly frustrating and disappointing seasons in the NFL, the one-time golden boy was no longer a hot commodity. But Bill Parcells, the master builder of the New York Jets, saw the untapped talent beyond Testaverde's dismal 48–83–1 record and signed him as backup insurance for Glenn Foley, the team's young and often injured starter. The policy began paying big dividends for Parcells after Foley incurred an early-season injury and Testaverde stepped in and led the Jets to a 10–1 record and the AFC Eastern Division championship. In what seemed like a sudden transformation but was actually a product of hard work, perseverance, and top-notch coaches and gifted teammates, Testaverde finally developed into the quarterback that everyone thought he would become when he first joined the NFL. "One thing I'm confident of now is that I can play this game as well as anyone," said Testaverde, after leading the AFC with a 101.6 quarterback rating and releasing the ghosts of past failures. "I don't worry about anything except what's happening to me right now. I'm not sure who said it, but I'm sort of living by this motto: Preoccupation with the past is the mental enemy of competitive excellence in the present."

CHARLES WOODSON

DEFENSIVE BACK • OAKLAND RAIDERS

CHARLES WOODSON

OAKLAND RAIDERS

Born: October 7, 1976, in Fremont, OH
Height: 6′ 1″ **Weight:** 200
College: Michigan **Round drafted:** First (1998)
NFL seasons: 1

CAREER STATS

Passes Defended	Int.	TDs	Tackles	Fumbles Forced	Fumbles Recovered
21	5	1	64	2	0

"Everything starts as a dream," said Charles Woodson on the night he was awarded the 1997 Heisman Trophy. For Woodson, who became the first defensive player ever to win the Heisman, the dream of winning the award for best collegiate player of the year had started when he was a young boy, playing ultimate football against his older brother, Terry, in the family living room in Fremont, Ohio. "You had two plays to score a touchdown," recalled Terry, noting that the football was a balled-up pair of socks and the end zone was the family couch. "Those were tough games. Anything went."

Woodson edged closer to his dream during his three seasons at Ross High School, where he played both offense and defense, fielded punts, and set the school's all-time career rushing and scoring records. In his senior year at Ross, Woodson set the school's single-season scoring and rushing records; averaged 19.5 yards per punt return; and recorded 63 tackles, three forced fumbles, four interceptions, and five deflections.

His heroics on the gridiron earned him the title Mr. Ohio Football for 1994, and a spot on the *USA Today* All-USA first team, as well as a scholarship to the University of Michigan. Deciding initially to concentrate on playing defense solely, Woodson played cornerback so spectacularly that he was named Big Ten Freshman of the Year and All-Big Ten First Team by the coaches in the conference. Woodson went national the following year, as he was named First Team All-American and the Chevrolet Defensive Player of the Year. The sophomore also lent some sock to the team's offense by racking up 152 yards on only six rushes, and 164 yards on thirteen pass receptions.

Woodson continued to sizzle in the Michigan secondary throughout his junior year, but he didn't clinch the Heisman Trophy until the final game of the regular season, when he put on a mind-boggling performance that sparked the top-ranked Wolverines to a big 20–14 win over Ohio State, which had come into the contest as the nation's fourth-ranked team. Woodson made the Buckeyes think that they were seeing triple, as he caught a 37-yard pass that set up Michigan's first touchdown, returned a punt 78 yards for a score, and made a touchdown-saving interception in the end zone that preserved Michigan's unbeaten season.

Snapped up by the Oakland Raiders with the fourth pick of the 1998 NFL draft, Woodson made an immediate impression on his new teammates. "He knew the Heisman didn't mean anything in the NFL, and that's what we like about him," said safety Anthony Newman. "The way you get respect here is by making plays—and that's just what he's been doing."

Woodson made so many plays that he was named the NFL's Defensive Rookie of the Year, and while he didn't get to play any offense, Raider coach Jon Gruden has already hinted that the role of his star cornerback will be expanding in the future. "He can give any unit of our football team a jolt," said Gruden. "When he gets his hands on the ball, he knows where to go and how to get there."

PEYTON MANNING

QUARTERBACK • INDIANAPOLIS COLTS

PEYTON MANNING

INDIANAPOLIS COLTS

Born: March 24, 1976, in New Orleans, LA
Height: 6′ 5″ **Weight:** 230
College: Tennessee **Round drafted:** First (1998)
NFL seasons: 1

CAREER STATS

Att.	Comp.	Pct.	Yards	TDs	Int.
575	326	56.7	3,739	26	28

A year before he graduated from the University of Tennessee, Peyton Manning knew, almost without a doubt, that he would be the number one selection of the 1998 NFL draft. "He's the first pick," declared Tom Boisture, the director of college scouting for the New York Giants. "Last year, this year, whenever he wants."

Manning—whose father, Archie, was a two-time college All-American and a 14-year NFL veteran—had, in a sense, been grooming himself to be a top-flight quarterback ever since he was 14 years old. Even at that early age, Manning was already preparing himself for the mental challenge of playing the position by analyzing films of NFL games. "If you're going to watch film, don't watch the ball," advised his father, who wouldn't allow his son to play organized tackle football until he was in the seventh grade. "Watch the defense. Fans watch the ball, quarterbacks watch the defense."

Although Manning obviously had some built-in advantages on his way to becoming a football superstar, he also brought along a fabulous work ethic that is all his own. "He came in with an attitude that I've never seen in any freshman," marveled former Tennessee teammate Eric Lane. "He wanted to get as much work done as possible, every day."

His talent and drive allowed him to become a four-year starter at Tennessee, where he set a stack of school, conference, and NCAA passing records, while leading the Vols to a 39–6 record and a top-ten national ranking in each of his last three seasons at the helm. "He's the best I've ever coached against, the best I've ever seen," raved Southern Mississippi defensive coordinator John Thompson. "You could watch TV all day (Sunday) and you won't see a better one."

Manning, as predicted, was the first overall pick of the 1998 draft, and he was instantly anointed as the lead pony in the Colts' corral. "The first day he came for mini camp, the weekend after we drafted him, he was the starting quarterback," declared Indianapolis head coach Jim Mora, who force-fed Manning a steady diet of hard-charging NFL defenses. "And I never got close to feeling that his confidence was low or that the team was down on him."

Manning, as expected, struggled through the early part of his rookie season while he learned to read complex NFL defenses and adjusted to playing against teams that are so much bigger, faster, and stronger than anything one sees on the college level. After six games, he had already been sacked 11 times and thrown 14 interceptions and only six touchdown passes. "There's no doubt he's still making mistakes, and sometimes it kills us," said Colts center Jay Leeuwenberg. "But you've got to make those type of mistakes. Everybody does when they're learning."

Manning showed just what an apt pupil he was over the remainder of the season, as he went on to finish among the league leaders in every major passing category while setting a pack of NFL rookie passing records, including ones in completions, attempts, touchdowns, and yardage. "He had a remarkable rookie season," said coach Mora. "He has the talent, the mental toughness, the attitude, and the character to be a special player in this league."

QUARTERBACK • MIAMI DOLPHINS

DAN MARINO

MIAMI DOLPHINS

Born: September 15, 1961, in Pittsburgh, PA
Height: 6' 3" **Weight:** 230
College: Pittsburgh **Round drafted:** First (1983)
NFL seasons: 16

CAREER STATS

Att.	Comp.	Pct.	Yards	TDs	Int.
7,989	4,763	59.6	58,913	408	235

Dan Marino, the last of six quarterbacks taken in the first round of the 1983 NFL draft, which is generally considered to have produced the single best crop of quarterbacks in NFL history, is also the last one standing. That famous class of 1983 included Jim Kelly, who helped pilot the Buffalo Bills to a record-setting four straight AFC Championships from 1990 to 1994, and the great John Elway, who quarterbacked the Denver Broncos through the wreckage of three Super Bowl losses before ending his career with a sweet pair of Super Sunday wins following the 1997 and 1998 seasons. Now, there's only Marino.

Although he was the last of the first-rounders to be selected, Marino was the first of the group to enjoy success. While Elway, for example, struggled through his early years in Denver, Marino became the only rookie to lead the AFC in passing, as well as the only rookie ever selected to start in the Pro Bowl. But Marino's first-year achievements were only a preamble for what he would do the following year, when he put on what might be the greatest quarterback show of all time.

In that second spectacular season, Marino shattered all previous single-season standards by passing for 5,084 yards and 48 touchdowns, records that still stand, not only unbroken, but never even seriously challenged. Marino carried his aerial circus into the postseason, where he added another 1,001 yards and eight touchdowns in three games, including a Super Bowl loss to a Joe Montana-led San Francisco 49ers team. "I find myself comparing him to John Unitas," said former Dolphins coach Don Shula, speaking about the Hall of Fame quarterback he coached in Baltimore. "And I didn't think I'd ever do that with any quarterback, much less a second-year one. But Dan makes such quick decisions, has such great peripheral vision, and the quickest arm I've ever seen."

Two years later, Marino showed that 1984 hadn't been a mirage by throwing 44 touchdown passes, the second-highest total in league history, and 4,746 yards, the third-highest total ever. "He's going to break all the records in the passing book," asserted Shula.

Marino has gone on to validate Shula's prediction by setting the career marks for completions (4,763), yards (58,913), and touchdowns (408), numbers that totally dwarf those of his nearest competitors. (Elway is second in both completions and yardage, with 4,123 and 51,475, respectively, while Fran Tarkenton, who retired in 1978, is the runner-up in touchdown passes, with 348.) The only jewel missing from his collection is a Super Bowl ring, and while winning the championship would cap his career like nothing else could, he refuses to allow its absence to cast a shadow.

"I'm not going to say it's not important," admitted Marino, who is about to begin his seventeenth season in the league, the lone remaining quarterback from the class of '83. "It would mean a lot to me. It's still my dream, and you have to have dreams. But if I don't get one, I'll still be able to look back knowing that I worked for that goal each and every year, and I don't think not winning a Super Bowl will detract from what I've accomplished."

JAKE PLUMMER

QUARTERBACK • ARIZONA CARDINALS

JAKE PLUMMER

ARIZONA CARDINALS

Born: December 19, 1974, in Boise, ID
Height: 6′ 2″ **Weight:** 197
College: Arizona State **Round drafted:** Second (1997)
NFL seasons: 2

CAREER STATS

Att.	Comp.	Pct.	Yards	TDs	Int.
843	481	57.1	5,940	32	35

As the youngest of three athletically active brothers, Jake Plummer learned early on how to play through adversity. "We've always competed in everything you could think of," remembered eldest brother Brett, who holds the Brown University 800-meter dash record. "And growing up, Jake never won at anything—*ever.*"

But little brother began rewriting the script at Capital High School, in Boise, Idaho, where he passed for 6,097 yards and 68 touchdowns in his last two prep school seasons. Plummer, who was a two-time all-state performer, is also an adept runner, and rushed for 645 yards and ten scores in his senior year. He found his next competitive challenge at Arizona State, where he announced his presence in dramatic style, by leading the Sun Devils on a 78-yard touchdown drive the very first time he quarterbacked the team. By the time he'd finished his four-year career, his dynamic style had made him the most popular player in ASU history, especially after he led the Sun Devils to 11 straight wins and the 1996 PAC-10 title.

Although he owned virtually all of ASU's career passing records, Plummer didn't have the size or arm strength that make most pro scouts drool. "When you saw him work out, you looked at his measurables and thought, he's fine, but nothing special," said the New York Giants' head coach, Jim Fassel. "Then you watched him in game-type situations, and he just made play after play."

Which is exactly what Plummer did on his first possession of his rookie year, after the Arizona Cardinals had selected him in the second round of the 1997 draft. In the fourth quarter of the team's sixth game, with one quarterback sidelined with an injury and the other yanked after putting up only three points in three quarters against the Philadelphia Eagles, Plummer engineered a 92-yard touchdown drive that put the Cards up, 10–7. "He trotted into the huddle and said, 'Come on guys, we can get it done,'" remembered wide receiver Frank Sanders. "We just kind of looked at each other. But he was the spark we needed. He just took command. Even as a rookie, he had it."

Plummer also closed out the 1997 season in memorable style, taking the Cards on an 80-yard scoring drive that put the team ahead 29–26 with .05 seconds left on the clock against the Atlanta Falcons. In between those signature drives, the Snake became only the fourth rookie to throw for four touchdowns in a game, and the first to rush for three touchdowns in a game since Peyton Manning's father, Archie, did it in 1977. But the season wasn't a nonstop joy ride for Plummer, who was constantly hounded and pounded, playing for a 4–12 team that had become resigned to losing. And 1998 didn't start out much better for the Cards, and Plummer, who was struggling to digest the new schemes installed by offensive coordinator Marc Trestman. But just when the Cards seemed stalled at 6–9, Plummer took them on an incredible journey, which included three straight come-from-behind wins, putting the Cardinals into the post-season for the first time since 1982. Then Plummer put he icing on the cake by leading Arizona to a huge upset over the Dallas Cowboys, in the franchise's first playoff victory since 1947. "He has unlimited potential," declared Marc Trestman. "He's a very special player. Jake's a natural leader, and a natural winner."

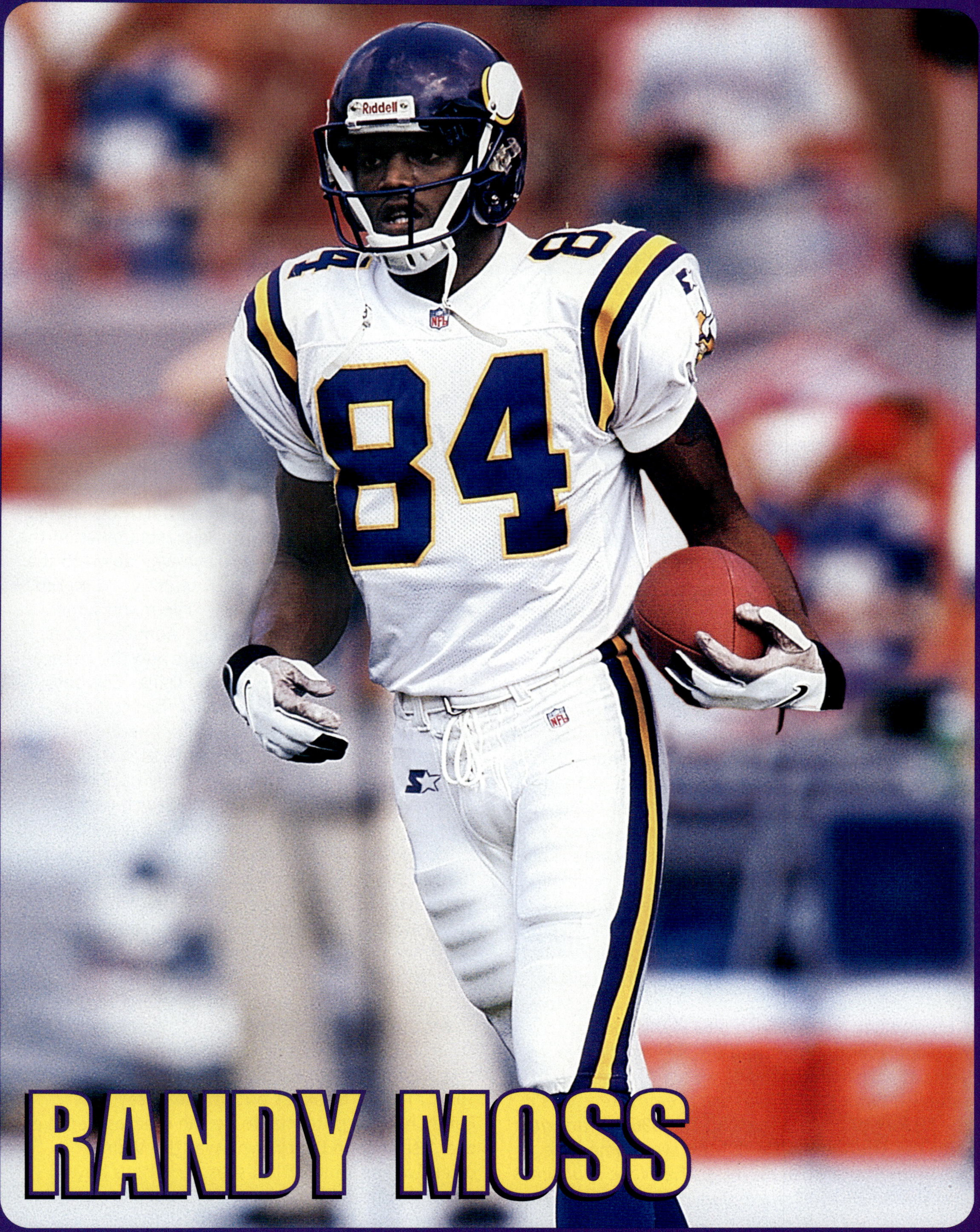

RANDY MOSS

WIDE RECEIVER • MINNESOTA VIKINGS

RANDY MOSS

MINNESOTA VIKINGS

Born: February 13, 1977, in Rand, WV
Height: 6' 4" **Weight:** 211
College: Marshall **Round drafted:** First (1998)
NFL seasons: 1

CAREER STATS

Receptions	Yards	Average	Long	TDs
69	1,313	19.0	61t	17

In a league of extraordinary athletes, Randy Moss, after only one season in the NFL, has already established himself as perhaps the most gifted of them all. "I've seen the best," said Minnesota Vikings kicker Gary Anderson, who has been in the league since 1982 and has played on teams with such exceptional receivers as Jerry Rice of the San Francisco 49ers, generally considered to be the greatest wide receiver of all time. "But I've never seen anything like this."

Moss has been drawing these types of raves ever since he was a four-sport star at DuPont High School in Rand, West Virginia. Moss—a high school teammate of Jayson Williams, the sensational rookie point guard of the NBA's Sacramento Kings—was such a gifted hoopster that he was twice named the West Virginia High School Basketball Player of the Year. Moss completed a rare double sweep as a senior, when he was also named the state's High School Football Player of the Year.

But Moss almost destroyed the yellow brick road that his talent and effort had built by committing a couple of irresponsible actions, which caused him to spend a month in jail and lose the college scholarship that Notre Dame had given him. Moss, who also had his scholarship to Florida State revoked, eventually wound up playing for two years at Marshall University, where he went on to establish himself as one of the greatest receivers in the history of college football. In 1996, his first year with the Thundering Herd, Moss was named the Southern Conference Player of the Week *four* times and snared 28 touchdown passes, tying an NCAA record first set by Jerry Rice in 1984. The following year, Moss added 23 TD grabs to his career total and was the winner of the Fred Biletnikoff award, which is given annually to the top receiver in college football.

With his credentials and potential, Moss should have been one of the top three players taken in the 1998 NFL draft, but he fell all the way down to the twenty-first pick, because most teams were scared off by the shadows in his past. "I've made some mistakes and I've paid the price," admitted Moss, who lost out on millions of dollars and the chance to play for his favorite team, the Dallas Cowboys. "I don't blame anyone but myself for what's happened to me. But I'm confident I'm headed in the right direction now. I realize what's at stake here. It's up to me to make the most of my opportunity."

The speedy receiver with the 39-inch vertical leap began exploiting his opportunity, and opposing defensive backs, in the season opener, when he toasted the Tampa Bay Buccaneers for a pair of long-distance touchdown receptions. "He's not the greatest receiver we've ever faced," declared Bucs defensive back Donnie Abraham, "but he will be."

Moss lent credence to Abraham's prediction by leading the league with a rookie-record 17 touchdown catches, leading to his being named the NFL Offensive Rookie of the Year. He also became the first freshman named to the All-Pro team since 1993. "I don't think it's out of line to say that Randy has the potential to be the greatest receiver ever," said Brian Billick, who was the Vikings' offensive coordinator last season. "That's how much talent he has."

JAMAL ANDERSON

RUNNING BACK • ATLANTA FALCONS

JAMAL ANDERSON

ATLANTA FALCONS

Born: September 30, 1972, in Newark, NJ
Height: 5′ 11″ **Weight:** 234
College: Utah **Round drafted:** Seventh (1994)
NFL seasons: 5

CAREER STATS

Rushing				Receiving			
Att.	Yards	Avg.	TDs	No.	Yards	Avg.	TDs
973	4,063	4.2	27	109	1,118	10.3	6

By the time he was nine years old, Jamal Anderson had started dominating football games and leveling defensive backs with his devastating straight-arm. Anderson graduated from Pop Warner to El Camino Real (California) High School, where he earned All-State honors as a senior in 1989. After spending a year at Moorpark Junior College, Anderson moved on to the University of Utah, where he produced two seasons of solid but unspectacular results. But after his second season, during which he ran for 1,030 yards and tacked on another 242 yards on 35 pass receptions, Anderson thought he'd be an early selection in the 1994 NFL draft.

To celebrate the big day, his family hosted a draft-day party and tuned the television to ESPN. But excitement turned to disappointment and then to embarrassment as Anderson fell all the way down to the seventh round, selected after 23 other running backs had already been chosen. "After the fifth round, I was like, 'I'm going out to play tennis,'" said Anderson, who was considered too slow to play tailback in the NFL, and too weak a blocker to play fullback.

When he arrived at the Atlanta Falcons training camp, Anderson noticed his name listed at the bottom of the team's depth chart at running back. Instead of wilting, he picked up a piece of chalk and drew an arrow from his name to the top of the list, boldly pointing in the direction that he intended to travel.

During his first season in Atlanta, Anderson mostly rode the bench, and in his second season his main contribution was as a kickoff returner. But by 1996, Anderson's hard work on the practice field and in the weight room, where he can bench-press 465 pounds, allowed him to take over as the team's featured back. Although he was at the top of the depth chart, he didn't gain too much notice, even after he had rushed for over 1,000 yards in two consecutive seasons. Heading into the 1998 season, in fact, one magazine listed him as only the twenty-second-best back in the NFL.

But Anderson emphatically rewrote that listing by galloping for 1,846 yards last season, the second-highest total in the league, behind only Terrell Davis of the Denver Broncos. Anderson, who was the NFC Offensive Player of the Month for November, was the main reason the Falcons were able to lead the league in time of possession and surprise virtually everyone by making it all the way to Super Bowl XXXIII. "I don't know if we've played against a back who is as determined and confident as Jamal," said Broncos defensive end Maa Tanuvasa. "He has spin moves and he can juke you, but he doesn't mind pinning his ears back and running over people. We understand that, like Terrell, he's a special player."

Anderson, who carried the ball 410 times, an NFL single-season record, isn't planning on slowing down anytime soon. "I've been getting pounded since I'm seven, but I don't feel it. I feel 100 percent. The way I figure it, I have seven or eight more years of dominance left in my body, and I won't leave anything on the field."

DEFENSIVE END • NEW YORK GIANTS

MICHAEL STRAHAN

NEW YORK GIANTS

Born: November 21, 1971, in Houston, TX
Height: 6′ 4″ **Weight:** 280
College: Texas Southern **Round drafted:** Second (1993)
NFL seasons: 6

CAREER STATS

Sacks	Tackles	Fumbles Forced	Fumbles Recovered	Int.	TDs
47.0	299	8	1	3	1

Michael Strahan, who moved to Mannheim, Germany, with his family when he was ten years old, has taken a most unlikely route to the NFL. Although he didn't have much of an opportunity to play organized football in Mannheim, his father had a strong hunch about his abilities. "I always had a feeling he had something special," said Gene Strahan. "He always had decent speed, and he was always dedicated. He'd always be telling his three older brothers, 'One day I'm going to be somebody. I'm going to be famous.' He has that kind of hunger."

Fortunately for Michael, he was able to move in his with his uncle Art, a former NFL player who lives in Texas, and play a year of high school football in a place where college coaches could evaluate his talent. Although he played well enough to win a scholarship to Texas Southern, he had no idea that he would ever emerge as an NFL star. "To be honest with you, I didn't really know what I was doing during my first couple of years of college ball," said Strahan. "And there were all these guys with huge muscles, who were faster and meaner than me. When pro scouts started coming to watch me play, it blew my mind."

The New York Giants saw enough potential to select Strahan in the second round of the 1993 draft. After struggling through a rookie year in which his development was slowed by injuries, Strahan strung together two solid seasons as the Giants' starter at right defensive end, before he was switched over to the left side prior to the start of the 1996 season. It was back to school for the unhappy Strahan, who struggled through a sub-par year while learning new footwork, techniques, and angles to the ball.

Although he had initially resisted the move, he ultimately embraced it, especially after John Fox was brought in as the defensive coordinator in 1997 and allowed his players to use their football instincts instead of being confined to a rigid, reactive system. Strahan immediately responded to the opportunity by recording a pair of sacks in the Giants' season opener. Making plays sent his confidence soaring, which lifted his playmaking to even higher levels. "When I get into a game, I *expect* to get a couple of sacks," said Strahan prior to the 1997 season. "Before, I was only *hoping* to get them. I've always worked hard, but I felt like I hadn't turned the corner yet. I used to wonder what it took to get there. I guess it takes persistence, a whole lot of hard work, and a coaching staff that believes in you. Eventually, you hope to have a year when you break out, and hopefully this will be my year."

Strahan got his wish in a big way, finishing the season among the league leaders with 14 sacks and winning selection to both the NFC Pro Bowl squad and the NFL All-Pro team. Showing that 1997 wasn't a mere fluke, Strahan then turned in a carbon-copy performance in 1998, solidifying his standing as one of the top defensive linemen in the game. "Michael is just such a all-around athlete," declared John Fox. "He not only stuffs the run, but he has more sacks over the last two seasons than anyone. He's something special."

GARRISON HEARST

RUNNING BACK • SAN FRANCISCO 49ERS

GARRISON HEARST

SAN FRANCISCO 49ERS

Born: January 4, 1971, in Lincolnton, GA
Height: 5′ 11″ **Weight:** 215
College: Georgia **Round drafted:** First (1993)
NFL seasons: 6

CAREER STATS

Rushing				Receiving			
Att.	Yards	Avg.	TDs	No.	Yards	Avg.	TDs
1,166	4,939	4.2	14	113	1,170	10.4	6

The last time Garrison Hearst lost a foot race was when he was 12 years old and came in second to his grandfather, Gyp Hearst, who was 65. "He was the fastest guy in Lincolnton," recalled Hearst, speaking about the small Georgia town where he was born and raised. "Yes, sir. He could roll. I guess he passed it down."

Hearst went on to have a high-voltage career at Lincoln County High School, where he threatened the all-time Georgia prep-school rushing record that the great Herschel Walker had set a decade earlier. "Garrison probably could have broken Walker's record," claimed Lincoln County High's assistant coach, Howard Ellis, who watched his prize running back rush for 2,097 yards and 36 touchdowns as a senior, while leading the Red Devils to a 15–0 record and the 1989 Class A state championship. "He was a joy to coach."

Hearst, who was named to the *USA Today* and *Parade Magazine* high school All-America teams, then moved on to the University of Georgia, where he was named co-winner of the Southeast Conference Freshman of the Year award. Two years later, in his third and final campaign with the Bulldogs, the swift and powerful runner set single-season SEC records for points scored (126), touchdowns (21), rushing touchdowns (190), and yards per carry (6.5). Hearst, who led the nation in touchdowns scored and who was the runner-up in rushing average per game (140.6), finished third in the balloting for the 1992 Heisman Trophy. "We've had some great running backs here, including Herschel Walker and Rodney Hampton," said former Bulldogs' coach Ray Goff. "But Garrison is the most complete football player I've ever seen at Georgia. Not only is he a great runner, but he's a great blocker and receiver, too."

Taken as the third overall pick of the 1993 draft by the Arizona Cardinals, Hearst was quickly sidetracked by a major knee injury, which limited him to just 14 games in his first two NFL seasons. But he bounced back big-time the following season, rushing for 1,070 yards and winning laurels as the NFL Comeback Player of the Year for 1995. Despite the award, Hearst was cut by the Cards and wound up in Cincinnati, where he became the Bengals' top rusher in 1996. Following that season, Hearst, who was a free agent, decided to sign with the San Francisco 49ers, even though they offered less money than the Bengals and Cards, who both wanted to resign him. "I love it here," said Hearst, who had grown weary of playing for losing teams. "Here they expect to win, which brings out the best in players."

Hearst has shown that he is a winner, too, by establishing himself as one of the most versatile backs in the league during his two seasons in the City by the Bay. After setting franchise records for longest run from scrimmage (96 yards) and rushing yards in a season (1,570) in 1998, Hearst is already drawing comparisons with 49ers icon Roger Craig, a great all-around back who helped spark the franchise to three Super Bowl wins in the 1980s. "He's the closest thing I've seen to Roger," said Tom Rathman, the team's running-back coach and former lead blocker for Craig. "He does it all, and with great quickness and speed."

ANTONIO FREEMAN

WIDE RECEIVER • GREEN BAY PACKERS

ANTONIO FREEMAN

GREEN BAY PACKERS

Born: May 27, 1972, in Baltimore, MD
Height: 6′ 1″ **Weight:** 198
College: Virginia Tech **Round drafted:** Third (1995)
NFL seasons: 4

CAREER STATS

Receptions	Yards	Average	Long	TDs
229	3,706	16.2	84t	36

Antonio Freeman started thinking about an NFL career in childhood, while starring on the playgrounds of Baltimore, Maryland. "I just somehow had a little more athletic talent than most of the other guys," said Freeman, who was an honorable mention on the *USA Today* All-America team as a high school senior and a three-year starter at Virginia Tech, where he set school records for career receptions (121) and receiving touchdowns (22). "Through Little League, high school, and college, I've always been very sure of my athletic abilities. I also worked hard on the field and in the classroom, studied, went to school, and listened to my parents. Everything I've accomplished stems from those sources."

Although Freeman realized a boyhood dream when he was drafted by the Green Bay Packers in 1995, he was stung at not being picked earlier than the third round. "I don't think I received the recognition or respect I deserved, because I wasn't a burner who could get down and run a 4.3 (seconds) 40 (yard dash)," said Freeman. "I came in feeling like I had something to prove. I was *extremely* hungry."

Freeman was also fortunate to have Robert Brooks, the Packers' outstanding wideout, as a teammate. "He passed on his words of wisdom (on how he had dealt with the situation two years earlier)," said Freeman, alluding to his teammate's rookie season. "I still thank him to this day for thinking enough of me, a complete stranger, to put that much work into me."

Freeman saw only limited playing time as a receiver in his rookie year, but he excelled at kickoff and punt returns, especially in the playoffs, when he was twice named NFL Special Teams Player of the Week. His performance in the postseason, which included a scintillating 76-yard punt return that broke open a close game against the Atlanta Falcons, boosted Freeman's confidence and opened the eyes of the Packers coaching staff.

The following year, Freeman worked his way into the starting lineup and quickly became the favorite target of quarterback Brett Favre. Just as he was hitting his stride, however, Freeman was sidelined for a month with a fractured arm. But he returned to the lineup and set a blistering pace over the final four games of the 1996 season. Despite wearing a cast on his left arm, Freeman managed to snare 25 passes for 452 yards and four touchdowns and was named the NFC Offensive Player of the Month. Freeman also again came up huge in postseason play, scoring a touchdown in each of the Packers' three games. His pivotal and record-setting 81-yard catch-and-run sparked Green Bay to a come-from-behind win against the New England Patriots in Super Bowl XXXI. "He took his game to one heck of a level," said Packers general manager Ron Wolff. "I expect a lot more great things from Antonio."

Freeman, who writes *Be the best* on his shoes each week, has fulfilled Wolff's great expectations by developing into one of the league's top receivers. "Each year I set my goals higher," said Freeman, who was named to the 1998 All-Pro team after a season in which he set career highs all across the board while leading the NFL with 1,424 receiving yards. "I just continue to work hard to become the best player I can be."

RUNNING BACK • DETROIT LIONS

BARRY SANDERS

DETROIT LIONS

Born: July 16, 1968, in Wichita, KS
Height: 5′ 8″ **Weight:** 203
College: Oklahoma State **Round drafted:** First (1989)
NFL seasons: 10

CAREER STATS

Rushing				Receiving			
Att.	Yards	Avg.	TDs	No.	Yards	Avg.	TDs
3,062	15,269	5.0	99	352	2,921	8.3	10

Barry Sanders has been rushing toward the Pro Football Hall of Fame since his first season with the Detroit Lions. Snagged with the third overall pick of the 1989 draft, Sanders set the tone of his record-breaking career in his first year by ripping off an NFC-best 1,470 yards and being named the NFL Offensive Rookie of the Year.

In a league in which running backs are frequently praised for rushing for 1,000 yards in a single season, Sanders has set entirely new standards by becoming the only running back in league history to top the 1,000-yard mark every year he's put on the uniform. With incredible and unmatched consistency, Sanders has returned to the gridiron each fall for ten consecutive seasons and been as dazzling as Michigan's woods ablaze with autumn's leaves. "Every game, just when you thought you'd seen it all, Barry does something that you've never seen before," said an admiring Dave Wannstedt, the assistant head coach of the Miami Dolphins.

Sanders sometimes seems as elusive as a phantom, slipping between defenders like a ghost through walls. Although he often makes lunging defenders look helpless, he never taunts an opposing player or adds an exclamation point to a touchdown by spiking the ball into the ground. "He doesn't do anything to bring individual attention to himself," noted Bobby Ross, who has been Detroit's head coach since 1997. "He has no special dance, he has no special wiggle. He just hands the ball to the official and trots off the field."

Even when defenses are able to stop Sanders for most of a game, their victory is usually temporary. "You can be doing really, really well," pointed out New York Giants, defensive coordinator John Fox. "And then—boom—he can ruin your day."

Which is exactly what Sanders did to the New York Jets in the final game of the 1997 season, a game that both teams needed to win in order to advance into the playoff picture. Through most of the third quarter, Sanders was limited to only 23 yards in 12 carries. Then he exploded for a 161 yards, including the touchdown burst that sent the Lions into the postseason. Sanders's late-game outburst gave him a whopping 2,053 rushing yards for the season, the second-highest total in NFL history. Finishing his remarkable season with an NFL-record 14 straight 100-yard rushing games, becoming the first runner in history to top 1,500 yards in five different seasons, Sanders shared league MVP honors with quarterback Brett Favre of the Green Bay Packers.

Sanders has been the most electrifying and productive back in the league during his ten years in the Motor City, and after a 1998 season in which he rushed for 1,491 yards, Sanders is poised to become the NFL's all-time leading rusher by the end of the 1999 campaign. With 15,269 yards, Sanders, who already ranks first or second in most rushing categories, is right on schedule to surpass the 16,726 rushing yards that the great Walter Payton amassed in his 13 seasons with the Chicago Bears. "I think everybody appreciates greatness," said Eric Dickerson, who rushed for 2,105 yards in 1984. "And that's what Barry Sanders is all about."

QUARTERBACK • MINNESOTA VIKINGS

RANDALL CUNNINGHAM

MINNESOTA VIKINGS

Born: March 27, 1963, in Santa Barbara, CA
Height: 6′ 4″ **Weight:** 234
College: UNLV **Round drafted:** Second (1985)
NFL seasons: 13

CAREER STATS

Att.	Comp.	Pct.	Yards	TDs	Int.
3,875	2,177	56.2	27,082	190	119

During his first tour of service in the NFL, Randall Cunningham established himself as one of the most gifted players in the league. "He's the best athlete to ever play the position," declared Buddy Ryan, who was the Eagles coach for five of the 11 years that Cunningham was in Philadelphia. "He has the best arm in the league. He's the best runner, and he could be the best punter."

Defensive coaches stayed up late trying to figure out ways to contain a quarterback who could hang the ball in the air for 80 yards and could also scramble and outrun many of the league's defensive backs. Cunningham, who has now run for more career yards than any other NFL quarterback, was such a willing and able scrambler that he rushed for more than 500 yards in five different seasons, including 1990, when he reached the astonishing total of 942. "You can't compare him to other quarterbacks," said veteran defensive coordinator Fritz Shurmur. "He's the one guy who can beat you on his own. He just creates something out of nothing."

Like a magician pulling rabbits out of his hat, Cunningham made the magical seem almost routine. But despite his extraordinary gifts, he never became a student of the game, never learned how to read defenses or to sit patiently in the pocket and pick out a secondary receiver. "I just used to fire the ball deep and hope somebody would catch it," recalled Cunningham, who decided to retire from football after the 1995 season, because he no longer liked the game or believed in his own ability.

After sitting out the 1996 season, working as a laborer, installing kitchen and bathroom countertops for a business that he owned, Cunningham decided to accept an offer from the Minnesota Vikings to back up the team's starter, Brad Johnson. "Getting up at six in the morning and filling a truck with tile and marble and granite will make you appreciate the other options in your life," he explained.

Although Cunningham did an adequate job of filling in when Johnson was sidelined at the tail end of the 1997 season, he was back on the bench at the start of the 1998 season. But during Minnesota's second game, Johnson went down again, and this time around, Cunningham was magnificent. Coming off the bench cold in the fourth quarter against the St. Louis Rams, with the score knotted at 31, Cunningham engineered a game-winning drive that he capped with a 19-yard touchdown pass to Cris Carter with 2:09 remaining in the game.

It was the start of a fabulous run for Cunningham, who went on to lead Minnesota to 13 wins in his 14 starts, as the Vikings became only the third team NFL history to post 15 wins in a season. Playing with the most control that he ever had, Cunningham threw for 3,704 yards, a career-high 34 touchdowns, only 10 interceptions, and a league-best 106.0 quarterback rating. "I think Randall is one of the great sports stories in a year that had some fabulous stories," said the Vikings' former offensive coordinator, Brian Billick, alluding to Cunningham's sudden transformation from being a selfish and undisciplined player to one who had become the calm and focused leader of the highest-scoring offense in NFL history. "It's a great story."

QUARTERBACKS

NAME	ATT.	COMP.	COMP%	YDs	YDs/ATT.	TDs	LONG	INT.	RATING
Randall Cunningham, MIN	425	259	60.9	3704	8.7	34	67t	10	106.0
Steve Young, SF	517	322	62.3	4170	8.1	36	81t	12	101.1
Chris Chandler, ATL	327	190	58.1	3154	9.7	25	78t	12	100.9
Troy Aikman, DAL	315	187	59.4	2330	7.4	12	67t	5	88.5
Steve Beuerlein, CAR	343	216	63.0	2613	7.6	17	68t	12	88.2
Brett Favre, GB	551	347	63.0	4212	7.6	31	84t	23	87.8
Charley Batch, DET	303	173	57.1	2178	7.2	11	98t	6	83.5
Eric Kramer, CHI	250	151	60.4	1823	7.3	9	79t	7	83.1
Trent Green, WAS	509	278	54.6	3441	6.8	23	75t	11	81.8
Jake Plummer, ARI	547	324	59.2	3737	6.8	17	57	20	75.0
Trent Dilfer, TB	429	225	52.4	2729	6.4	21	79t	15	74.0
Troy Banks, STL	408	241	59.1	2535	6.2	7	80t	14	68.6
Danny Kanell, NYG	299	160	53.5	1603	5.4	11	46	10	67.3
Kerry Collins, NO	353	170	48.2	2213	6.3	12	89t	15	62.0
Bobby Hoying, PHI	224	114	50.9	961	4.3	0	38	9	45.6

RUSHERS

NAME	ATT.	YDs	YDs/ATT.	LONG	TDs
Jamal Anderson, ATL	410	1846	4.5	48	14
Garrison Hearst, SF	310	1570	5.1	96t	7
Barry Sanders, DET	343	1491	4.3	73t	4
Emmitt Smith, DAL	319	1332	4.2	32	13
Robert Smith, MIN	249	1187	4.8	74t	6
Duce Staley, PHI	258	1065	4.1	64t	5
Gary Brown, NYG	247	1063	4.3	45	5
Adrian Murrell, ARI	274	1042	3.8	32	8
Warrick Dunn, TB	245	1026	4.2	50	2
Mike Alstott, TB	215	846	3.9	37	8

RECEIVERS

NAME	REC.	YDs	YDs/REC.	LONG	TDs
Frank Sanders, ARI	89	1145	12.9	42	3
Antonio Freeman, GB	84	1424	17.0	84t	14
Herman Moore, DET	82	983	12.0	36	5
Jerry Rice, SF	82	1157	14.1	75t	9
Cris Carter, MIN	78	1011	13.0	54t	12
Michael Irvin, DAL	74	1057	14.3	51	1
Larry Centers, ARI	69	559	8.1	54	2
Raghib Ismail, CAR	69	1024	14.8	62	8
Johnnie Morton, DET	69	1028	14.9	98t	2
Randy Moss, MIN	69	1313	19.0	61t	17

PUNTERS

NAME	PUNTS	YDs	LONG	YDs/PUNT
Mark Royals, NO	88	4017	64	45.6
Brad Maynard, NYG	101	4566	63	45.2
Mitch Berger, MIN	55	2458	67	44.7
Rick Tuten, STL	95	4202	64	44.2
Matt Turk, WAS	93	4103	69	44.1
John Jett, DET	66	2892	60	43.8
Toby Gowin, DAL	77	3342	65	43.4
Sean Landeta, GB	65	2788	72	42.9
Reggie Roby, SF	60	2511	66	41.9
Tom Hutton, PHI	104	4339	61	41.7
Scott Player, ARI	81	3378	67	41.7
Mike Horan, CHI	64	2643	57	41.3
Tommy Barnhardt, TB	81	3340	55	41.2
Ken Walter, CAR	77	3131	59	40.7
Dan Stryzinski, ATL	74	2963	55	40.0

PUNT RETURNERS

NAME	RET.	YDs	YDs/RET.	LONG	TDs
Deion Sanders, DAL	24	375	15.6	69t	2
Jacquez Green, TB	30	453	15.1	95t	1
Andre Hastings, NO	22	307	14.0	76	0
Glyn Milburn, CHI	25	291	11.6	93t	1
Brian Mitchell, WAS	44	506	11.5	47	0
Winslow Oliver, CAR	44	464	10.5	35	0
Eddie Kennison, STL	40	415	10.4	71t	1
David Palmer, MIN	28	289	10.3	53	0
Roell Preston, GB	44	398	9.0	71t	1
R.W. McQuarters, SF	47	406	8.6	72t	1

KICKOFF RETURNERS

NAME	RET.	YDs	YDs/RET.	LONG	TDs
Terry Fair, DET	51	1428	28.0	105t	2
Tim Dwight, ATL	36	973	27.0	93t	1
Roell Preston, GB	57	1497	26.3	101t	2
Michael Bates, CAR	59	1480	25.1	99t	1
Glyn Milburn, CHI	62	1550	25.0	94t	2
Kevin Mathis, DAL	25	621	24.8	42	0
Allen Rossum, PHI	44	1080	24.5	54	0
Reidel Anthony, TB	46	1118	24.3	60	0
David Palmer, MIN	50	1176	23.5	88t	1
Tony Horne, STL	56	1306	23.3	102t	1

TOUCHDOWNS

NAME	TDs	RUSHING	RECEIVING	RETURNS	POINTS
Randy Moss, MIN	17	0	17	0	106*
Jamal Anderson, ATL	16	14	2	0	98*
Terrell Owens, SF	15	1	14	0	92*
Emmitt Smith, DAL	15	13	2	0	90
Antonio Freeman, GB	14	0	14	0	86*
Cris Carter, MIN	12	0	12	0	72
Terance Mathis, ATL	11	0	11	0	66
Leroy Hoard, MIN	10	9	1	0	60
Adrian Murrell, ARI	10	8	2	0	60
Mike Alstott, TB	9	8	1	0	54
Garrison Hearst, SF	9	7	2	0	56*
Jerry Rice, SF	9	0	9	0	58*
Leslie Shepherd, WAS	9	1	8	0	56*
Skip Hicks, WAS	8	8	0	0	48
Raghib Ismail, CAR	8	0	8	0	48
Robert Smith, MIN	8	6	2	0	48
J.J. Stokes, SF	8	0	8	0	48

** includes 2-point conversions*

KICKERS

NAME	PAT	ATT.	FG	ATT.	LONG	POINTS
Gary Anderson, MIN	59	59	35	35	53	164
Ryan Longwell, GB	41	43	29	33	45	128
Richie Cunningham, DAL	40	40	29	35	54	127
Morten Andersen, ATL	51	52	23	28	53	120
Jason Hanson, DET	27	29	29	33	51	114
Wade Richey, SF	49	51	18	27	46	103
Brad Daluiso, NYG	32	32	21	27	51	95
Michael Husted, TB	29	30	21	28	52	92
John Kasay, CAR	35	37	19	26	56	92
Doug Brien, NO	31	31	20	22	56	91
Jeff Jaeger, CHI	27	28	21	26	52	90
Jeff Wilkins, STL	25	26	20	26	57	85
Joe Nedney, ARI	30	30	13	19	53	69
Cary Blanchard, WAS	30	31	11	17	54	63

TEAM OFFENSE

NAME	RUSH YARDS	PASS YARDS	TOTAL YARDS	RUSH TDs	PASS TDs	RET.& REC.TDs	EP/ 2PT.CON	FG/ FGA	POINTS
Minnesota	1936	4492	6428	17	41	6	59/3	35/35	556
San Francisco	2544	4510	7054	19	41	1	49/5	18/27	479
Atlanta	2101	3744	5845	18	28	7	51/1	23/28	442
Green Bay	1526	4340	5866	7	33	6	41/1	29/33	408
Dallas	2014	3546	5560	21	17	4	40/0	29/35	381
Carolina	1458	3624	5082	11	25	4	35/2	19/26	336
Arizona	1627	3768	5395	18	17	1	36/0	23/33	325
Washington	1685	3724	5409	15	24	1	36/1	13/23	319
Tampa Bay	2148	2787	4935	12	21	3	29/3	21/28	314
Detroit	1955	3398	5353	12	17	3	27/0	29/33	306
New Orleans	1325	3514	4839	6	19	10	31/0	20/22	305
N.Y. Giants	1889	2822	4711	10	18	4	32/0	21/27	287
St. Louis	1385	3381	4766	17	12	3	25/4	20/26	285
Chicago	1713	3277	4990	9	16	5	27/1	21/26	276
Philadelphia	1775	2730	4505	10	7	0	15/0	14/21	161

SACKS

NAME	SACKS
Reggie White, GB	16.0
Chris Doleman, SF	15.0
Kevin Greene, CAR	15.0
Michael Strahan, NYG	15.0
Hugh Douglas, PHI	12.5
Roy Barker, SF	12.0
Kevin Carter, STL	12.0
Robert Porcher, DET	11.5
Chad Bratzke, NYG	11.0
John Randle, MIN	10.5
Lester Archambeau, ATL	10.0
La'Roi Glover, NO	10.0
Simeon Rice, ARI	10.0
Bryant Young, SF	9.5
Brad Culpepper, TB	9.0
Mark Smith, ARI	9.0

INTERCEPTIONS

NAME	INT.	YDs	LONG	TDs
Kwamie Lassiter, ARI	8	80	29	0
Ray Buchanan, ATL	7	102	34	0
Jimmy Hitchcock, MIN	7	242	79t	3
Sammy Knight, NO	6	171	91t	2
Eric Davis, CAR	5	81	56t	2
Percy Ellsworth, NYG	5	92	43t	2
Robert Griffith, MIN	5	25	17	0
Ronald McKinnon, ARI	5	25	17	0
Deion Sanders, DAL	5	153	71t	1
Tyrone Williams, GB	5	40	15	0
12 players tied with	4			

TEAM DEFENSE

NAME	RUSH YARDS	PASS YARDS	TOTAL YARDS	OPP. TDs	RUSH TDs	PASS TDs	RET. TDs	SACKS	INT.	POINTS ALLOWED
Dallas	1619	3545	5164	32	10	21	1	34	14	275
Atlanta	1203	3531	4734	35	7	22	6	38	19	289
Tampa Bay	1583	2762	4345	31	12	15	4	37	12	295
Minnesota	1614	3452	5066	35	12	17	6	38	19	296
N.Y. Giants	2004	3167	5171	34	13	17	4	54	19	309
Green Bay	1442	3065	4507	36	7	23	6	50	13	319
San Francisco	1610	3733	5343	39	13	25	1	51	21	328
Philadelphia	2416	2720	5136	39	18	18	3	42	9	344
New Orleans	1700	3968	5668	40	13	24	3	47	21	359
Chicago	1875	3228	5103	42	12	27	3	28	14	368
St. Louis	2049	2831	4880	43	11	28	4	50	16	378
Detroit	2102	3015	5117	43	15	23	5	43	12	378
Arizona	1989	3276	5265	44	18	21	5	39	20	378
Carolina	2133	3709	5842	46	14	30	2	37	19	413
Washington	2436	2918	5354	50	24	21	5	33	13	421

OFFENSE

QUARTERBACKS

Steve Young-x	49ers
Chris Chandler	Falcons
Randall Cunningham	Vikings

RUNNING BACKS

Barry Sanders-xy	Lions
Jamal Anderson-x	Falcons
Garrison Hearst-y	49ers
Emmitt Smith-z	Cowboys
Robert Smith-z	Vikings

FULLBACK

Mike Alstott	Buccaneers

WIDE RECEIVERS

Antonio Freeman-x	Packers
Randy Moss-x	Vikings
Cris Carter	Vikings
Jerry Rice	49ers

TIGHT ENDS

Mark Chmura-x	Packers
Wesley Walls	Panthers

TACKLES

Larry Allen-x	Cowboys
William Roaf-xy	Saints
Todd Steussie	Vikings
Bob Whitfield-z	Falcons

GUARDS

Randall McDaniel-x	Viking
Kevin Gogan-x	49ers
Nate Newton	Cowboys

CENTERS

Jeff Christy-x	Vikings
Tony Mayberry	Buccaneers

DEFENSE

DEFENSIVE ENDS

Reggie White-x	Packers
Michael Strahan-x	Giants
Joe Johnson	Saints

INTERIOR LINEMEN

John Randle-x	Vikings
Warren Sapp-x	Buccaneers
Leon Lett	Cowboys

OUTSIDE LINEBACKERS

Derrick Brooks-x	Buccaneers
Kevin Greene-x	Panthers
Jessie Armstead	Giants

INSIDE LINEBACKERS

Jessie Tuggle-x	Falcons
Ed McDaniel-y	Vikings
Hardy Nickerson-z	Buccaneers
Winfred Tubbs-z	49ers

CORNERBACKS

Deion Sanders-x	Cowboys
Ray Buchanan-x	Falcons
Aeneas Williams	Cardinals

STRONG SAFETIES

LeRoy Butler-x	Packers
Darren Woodson	Cowboys

FREE SAFETY

Eugene Robinson-x	Falcons

SPECIALISTS

PUNTER

Matt Turk	Redskins

KICKER

Gary Anderson	Vikings

KICK RETURNER

Roell Preston	Packers

SPECIAL TEAMS

Michael Bates	Panthers

NOTES: x - starter; y - injured; z - injury replacement;
n - "need player," selected by coach from
first alternates at defensive end, inside linebacker, or outside linebacker

PREDICTIONS FOR THE 1999 SEASON

NATIONAL FOOTBALL CONFERENCE

EAST	**CENTRAL**	**WEST**
Arizona	Minnesota	Atlanta
Dallas	Green Bay	San Francisco
Washington	Tampa Bay	St. Louis
N.Y. Giants	Detroit	New Orleans
Philadelphia	Chicago	Carolina

NFC Champion: Minnesota Vikings

AMERICAN FOOTBALL CONFERENCE

EAST	**CENTRAL**	**WEST**
N.Y. Jets	Jacksonville	Denver
Miami	Baltimore	Seattle
New England	Pittsburgh	Kansas City
Buffalo	Tennessee	Oakland
Indianapolis	Cleveland	San Diego
	Cincinnati	

AFC Champion: Denver Broncos

SUPER BOWL WINNER: Denver Broncos

CHALLENGE THE EXPERT

Now it's time for you to sharpen up your pencil and put on your thinking cap.

NATIONAL FOOTBALL CONFERENCE

EAST	CENTRAL	WEST
1. ______	1. ______	1. ______
2. ______	2. ______	2. ______
3. ______	3. ______	3. ______
4. ______	4. ______	4. ______
5. ______	5. ______	5. ______

NFC Champion: ______________

AMERICAN FOOTBALL CONFERENCE

EAST	CENTRAL	WEST
1. ______	1. ______	1. ______
2. ______	2. ______	2. ______
3. ______	3. ______	3. ______
4. ______	4. ______	4. ______
5. ______	5. ______	5. ______
	6. ______	

AFC Champion: ______________

SUPER BOWL WINNER: ______________

Although I've filled this book with stories and stats about athletic accomplishments, we shouldn't lose sight of the fact that football is a game, and what the players do when they're not playing is probably a lot more meaningful than what they do on the field. Playing football is what these players do as their job—what they do with the rest of their time, with their families and in their communities, is what defines them as people. All of the players I've written about in this book have chosen to give their time and their money to people and causes that are in need of their help (if they hadn't, they wouldn't be in the book). Here are some of the things that the players have done.

Terrell Davis has been a generous contributor to the sports programs at his former high school, donating lots of equipment and taking the time to go back to the school and encourage its student body to rise above their impoverished circumstances and strive for success. **Drew Bledsoe** has created his own foundation to improve the lives of children by teaching adults better parenting skills, and he has also endowed a scholarship at Washington State University. **Doug Flutie** formed a foundation to raise funds and create awareness for autistic children. A portion of all proceeds from the sale of the cereal named after him, Flutie Flakes, goes directly to the foundation. **Mark Brunell** hosts an annual golf tournament to benefit Wolfson's Children Hospital and frequently visits with children who are facing life-threatening diseases. **Vinny Testaverde** has done volunteer work for an organization called Literacy Work. **Charles Woodson** has helped raise money for the Children's Hospital of Oakland, and he also helped to feed 120 single-parent families for Thanksgiving. **Peyton Manning** serves as the national spokesperson for C.A.S.A., a program that represents children in juvenile court cases, and he also coached a flag football game between underprivileged children to raise money for the program. **Dan Marino** has been as great off the field as he has been on it, donating huge sums of money and time to benefit thousands of children through the Dan Marino Foundation. Marino, who is responsible for the building of Miami Children's Hospital/Dan Marino Center in Weston, Florida, was named the 1998 Sprint/NFL Man of the Year for his service to the community. **Jake Plummer** has set up a foundation to benefit people with Alzheimer's disease, and he has also made plans to upgrade the athletic facility at his former high school, as well as to create a scholarship at Arizona State. **Randy Moss** has created Randy's Purple Pioneers, which, among other things, gives out free tickets to Vikings home games to children who have altered their behavior in a positive direction. **Jamal Anderson** enjoys talking to student groups and creating a positive impact on children. After reading an article about a boy who gave up his football career by donating one of his kidneys to save his grandmother's life, Anderson flew the boy into Atlanta as his guest for a weekend, which included a game against the St. Louis Rams. Anderson was also the team's spokesperson for the 1998 Tackle Your Hunger Campaign, the Falcons' annual holiday canned food drive. **Michael Strahan** is deeply involved with H.E.L.P., the nation's largest provider of housing and services for the homeless. "I've always liked helping people," said Strahan, who also provides scholarships to the five-day football camp that he hosts. "When you're good to people, it comes back to you tenfold. It's actually fun to raise money and awareness and see the impact that you have on people who need it." **Garrison Hearst** sponsors children's sports teams in his home town of Lincolnton, Georgia. **Antonio Freeman** sponsored a basketball team in Baltimore's Oliver Recreation League, where he played as a boy. **Barry Sanders** has been generous with many groups, including the P.A.Y.-by-Play program, which he created to benefit the Special Olympics. **Randall Cunningham** donated $100,000 to build a school gym dedicated in memory of Jerome Brown, a former Eagles teammate who was killed in a car accident in 1992.

Maybe you would like to do something to help people in your community. There are plenty of wonderful causes for which you can volunteer your time and make a real impact on people's lives. Check with your parents or your school, if you want some suggestions. Maybe there are things to be done right in your own home. If every day we all do something to help make the world a kinder, safer, cleaner place, then we'll all start living on a kinder, safer, cleaner planet.

One thing that we can all start doing immediately is to smile at the people we meet, and to never pick on or denigrate another human being.

Richard J. Brenner

ORDER FORM

If you enjoyed this book, you might want to order some of our other exciting titles written by Richard J. Brenner, the best-selling sportswriter in America.

TITLE	QTY.
BASEBALL SUPERSTARS ALBUM 1999 Includes 16 full-color pages of the game's top players, plus career stats. *48 pages. ($4.50/$6.50 Can.)*	______
BASKETBALL SUPERSTARS ALBUM 1999 Includes 16 full-color pages of the game's top players, plus career stats. *48 pages. ($4.50/$6.50 Can.)*	______
FOOTBALL SUPERSTARS ALBUM 1999 Includes 16 full-color pages of the game's top players, plus career stats. *48 pages. ($4.99/$6.99 Can.)*	______
MARK McGWIRE An easy-to-read, photo-filled biography especially for young readers. *32 pages. ($4.50/$6.50 Can.)*	______
SAMMY SOSA An easy-to-read, photo-filled biography especially for young readers. *32 pages. ($4.50/$6.50 Can.)*	______
MICHAEL JORDAN An easy-to-read, photo-filled biography especially for young readers. *32 pages. ($3.95/$5.95 Can.)*	______
KOBE BRYANT An easy-to-read, photo-filled biography especially for young readers. *32 pages. ($3.95/$5.95 Can.)*	______
GRANT HILL An easy-to-read, photo-filled biography especially for young readers. *32 pages. ($3.50/$4.50 Can.)*	______
DEION SANDERS * BRETT FAVRE A dual biography of two All-Pros. *96 pages, incl. 10 pages of photos. ($3.99/$5.00 Can.)*	______
STEVE YOUNG * JERRY RICE A dual biography of two All-Pro teammates. *112 pages, incl. 10 pages of photos. ($3.99/$5.50 Can.)*	______
THE WORLD SERIES, THE GREAT CONTESTS Seven of the most exciting of the Fall Classics are brought to life. *176 pages, incl. 16 action-packed photos. ($4.50/$5.50 Can.)*	______
SUPER SPORTS STARS Up close with 10 of your favorite superstars, including Mark McGwire, Terrell Davis, Shaquille O'Neal, and Eric Lindros. *64 pages, incl. 10 pages of photos. ($3.99/$5.99 Can.)*	______
SUPER SPORTS STARS BOOK OF RECORDS, by Bob Carroll and Pete Palmer. Career, game, and season records for baseball, football, basketball, and hockey. *64 pages. ($3.99/$5.99 Can.)*	______
TOTAL NUMBER OF BOOKS ORDERED	______
POSTAGE AND HANDLING *($1.50 per book, up to a maximum of $7.50; $1.75, up to a maximum of $8.75, in Canada.)*	______
TOTAL PAYMENT ENCLOSED *(Payment must accompany all orders. All payments must be in U.S. dollars.)*	______

NAME ______________________________

ADDRESS ______________________________

CITY ______________ **STATE** ______ **ZIP CODE** ______ **COUNTRY** ______

SEND PAYMENT TO: EAST END PUBLISHING, 54 ALEXANDER DR., SYOSSET, NY 11791
Discounts are available for orders of 25 or more books. For details, write or call (516) 364-6383.